Title: "The Precious Blue Sapphire Gemstone and Its Global Value"

Warm regards,

FAISAL JAMIL

I Always Give's Free Copies Need Your Feedback And

Reviews Keeps In Touch!

http://www.amazon.com/author/faisal.jamil

Email: faisaljamilauthor@gmail.com

About the author

Certainly! Faisal Jamil is a multifaceted individual with a diverse set of skills and experiences. With a strong foundation in computer knowledge since childhood, he has developed a deep understanding of technology that informs his work as a content writer. Faisal also possesses digital skills, which further enhance his abilities in various digital platforms and technologies.

Beyond his professional endeavors, Faisal Jamil has also excelled in the martial arts, particularly Shotokan Karate, where he achieved the prestigious rank of first Dan black belt. This achievement speaks to his dedication, discipline, and commitment to personal growth and mastery.

In his professional life, Faisal Jamil has carved out a successful career in sales management within the Fast Moving Consumer Goods (FMCG) sector. His roles in various FMCG companies have honed his skills in strategic planning, team leadership, and business development. Faisal's ability to drive sales and achieve targets has been instrumental in his career progression, showcasing his talent for identifying opportunities and delivering results.

Faisal Jamil is also deeply interested in business investment strategies, planning, and execution. His understanding of these areas has been key to his success in the business world, allowing him to make informed decisions and implement effective strategies. His ability to navigate the complexities of investment planning and execution has set him apart as a strategic thinker and a valuable asset in any business endeavor.

Overall, Faisal Jamil is a dynamic individual who combines his passion for technology, martial arts, sales management, digital skills, and business investment strategies to achieve success in diverse fields. His journey is a testament to his versatility, resilience, and continuous pursuit of excellence.

Yours Sincerely

FAISAL JAMIL

I Always Give's Free Copies Need Your Feedback And

Reviews Keeps In Touch!

https://www.amazon.com/author/faisal.jamil

Email: faisaljamilauthor@gmail.com

THE PRECIOUS

BLUE SAPPHIRE

GEMSTONE

AND ITS GLOBAL VALUE

Table of Content

Preface

Blue sapphires have captivated the hearts and minds of humanity for centuries. Their deep, mesmerizing blue hues symbolize not just beauty, but also a rich history and profound cultural significance. "The Precious Blue Sapphire Gemstone and Its Global Value" is born from a passion for these remarkable gems and a desire to share their story with enthusiasts, collectors, and anyone drawn to their allure.

The journey of a blue sapphire from deep within the Earth to an exquisite piece of jewelry is a tale of natural wonder, human ingenuity, and the intersection of science, art, and commerce. This book aims to take readers on this journey, exploring every facet of blue sapphires, from their geological formation and historical significance to their contemporary market value and future prospects.

In **Chapter 1**, we begin with the origins and history of blue sapphires. This chapter provides a foundation for understanding the gemstone's allure by delving into its composition, legendary stories, and historical artifacts that have featured these magnificent stones.

Chapter 2 takes us deep into the Earth to uncover the geological processes that form blue sapphires. By understanding where and how these gemstones are formed, readers gain insight into their rarity and value. This chapter also highlights the major sapphire deposits around the world and discusses the methods used to extract these gems responsibly.

The focus of **Chapter 3** is on the value and grading of blue sapphires. Through a detailed examination of the four Cs—Color, Clarity, Cut, and Carat weight—this chapter demystifies the criteria that determine a sapphire's worth. It also explores the grading systems used by leading gemological institutes and examines current trends in the gemstone market.

Chapter 4 shifts the spotlight to the cultural and artistic significance of blue sapphires. From their symbolic meanings in different cultures to their use in iconic jewelry pieces, this chapter celebrates the artistry involved in crafting sapphire jewelry and the enduring appeal of these gemstones in fashion and tradition.

Finally, **Chapter 5** looks towards the future of blue sapphires, addressing the industry's move towards sustainable and ethical practices. It explores the technological advancements that are revolutionizing the mining, cutting, and treatment of sapphires. Additionally, this chapter provides valuable insights for collectors and investors, highlighting the importance of authentication, provenance, and proper care.

The creation of this book has been a journey of discovery and admiration for the blue sapphire. It is my hope that readers will find as much fascination and inspiration in these pages as I have found in researching and writing them. Whether you are a seasoned gemologist, a collector, or simply someone enchanted by the beauty of blue sapphires, this book is designed to offer a comprehensive and engaging exploration of these precious gemstones.

I invite you to turn the pages and embark on this journey with me, to uncover the splendor and significance of blue sapphires, and to appreciate their timeless value in our world.

Sincerely,

Faisal Jamil

INTRODUCTION

Blue sapphires, with their captivating allure and deep, mesmerizing hues, have long been revered as some of the most precious gemstones in the world. Their beauty is matched only by their rich history and profound cultural significance. "The Precious Blue Sapphire Gemstone and Its Global Value" is a comprehensive exploration of these remarkable stones, designed to illuminate their journey from the depths of the Earth to the most exquisite pieces of jewelry and beyond.

This book aims to provide readers with an in-depth understanding of blue sapphires, examining every aspect of these gemstones—from their geological formation and historical importance to their market value and future prospects. Whether you are a gemstone enthusiast, a collector, an investor, or simply someone fascinated by the natural world, this book will offer valuable insights and knowledge about blue sapphires.

The Origins and History of Blue Sapphires

Our journey begins with an exploration of the origins and history of blue sapphires. Understanding where these gemstones come from and how they have been perceived throughout history provides a foundation for appreciating their value. We will delve into ancient myths and legends, the role of blue sapphires in various cultures, and the famous sapphires that have adorned the crowns and jewelry of royalty.

The Geological Formation of Blue Sapphires

To truly appreciate the rarity and value of blue sapphires, it is essential to understand the natural processes that create them. We will explore the high-pressure and high-temperature conditions deep within the Earth that give rise to these gems. Additionally, this section will highlight the major sapphire deposits around the world and the unique geological conditions that contribute to the distinct characteristics of sapphires from different regions.

The Value and Grading of Blue Sapphires

Determining the value of blue sapphires involves a complex interplay of various factors. This book will provide a detailed examination of the four Cs—Color, Clarity, Cut, and Carat weight—that are crucial in grading these gemstones. By understanding these criteria, readers will be better equipped to evaluate sapphires and make informed decisions when buying or investing in them. We will also look at the grading systems used by leading gemological institutes and analyze current market trends.

Blue Sapphires in Jewelry and Culture

Blue sapphires are not only valuable for their intrinsic properties but also for their cultural and artistic significance. We will explore how these gemstones have been used in symbolic jewelry, their meanings in different cultures, and their role in iconic pieces such as engagement rings and royal regalia. This section will also highlight the craftsmanship involved in creating sapphire jewelry, showcasing the work of notable designers and jewelers.

The Future of Blue Sapphires

As we look to the future, sustainability and ethical practices in the gemstone industry are becoming increasingly important. This book will discuss the efforts to make sapphire mining more sustainable and the role of fair trade practices in ensuring ethical sourcing. We will also explore technological advancements that are transforming the industry, including innovations in mining, cutting, and treatment processes. Additionally, this section will provide insights into blue sapphires as an investment, offering tips on building and maintaining a valuable collection.

Conclusion

"The Precious Blue Sapphire Gemstone and Its Global Value" is a tribute to the enduring appeal and significance of blue sapphires. By delving into the rich tapestry of their history, formation, value, and cultural impact, this book aims to offer a comprehensive and engaging exploration of these precious gemstones. Whether you are a seasoned gemologist or a curious reader, I hope this book will deepen your appreciation for blue sapphires and the remarkable journey they undertake from the heart of the Earth to the pinnacle of human artistry.

Join me in uncovering the splendor and significance of blue sapphires, and in celebrating their timeless value in our world.

Faisal Jamil

Chapter 1
The Origins and History of Blue Sapphires

Introduction to Blue Sapphires

Blue sapphires, with their enchanting deep blue hues, have been objects of fascination and admiration for countless generations. These precious gemstones are part of the corundum family, sharing a close relationship with rubies. The distinct blue color of sapphires is primarily due to the presence of trace elements such as titanium and iron within the crystal structure. The interplay of these elements in the corundum matrix creates a spectrum of blue shades, from pale sky blue to the most coveted deep, velvety blue known as "cornflower blue."

The allure of blue sapphires is not just in their visual appeal but also in their inherent physical properties. Corundum is one of the hardest minerals on the Mohs scale, second only to diamonds, which makes sapphires highly durable and ideal for use in various types of jewelry. This combination of beauty and resilience has ensured that blue sapphires remain among the most sought-after gemstones in the world.

Historical Significance

The history of blue sapphires is as rich and varied as the cultures that have treasured them. These gemstones have played significant roles in myths, religions, and royal traditions across the globe.

Ancient Persia

In ancient Persia, blue sapphires were believed to be pieces of the sky that had fallen to earth. The Persians thought that the earth rested on a giant blue sapphire, and its reflection was what gave the sky its blue color. This belief highlighted the deep reverence and mystical significance they attached to these gemstones.

Medieval Europe

During the medieval period in Europe, blue sapphires were considered symbols of Heaven. Clergy members wore sapphires as they believed the stones would help them connect with the divine and guard against envy and harm. The blue sapphire's association with celestial and spiritual purity made it a favored gemstone among the religious elite.

Royal Adornments

Throughout history, blue sapphires have been integral to royal regalia. Kings and queens adorned their crowns, scepters, and jewelry with these gems, which were believed to bestow wisdom, virtue, and good fortune. The association of blue sapphires with royalty and nobility further cemented their status as gemstones of great importance and value.

Symbolism and Legends

Blue sapphires have also been attributed with various mystical properties. They were believed to protect their wearers from envy and harm, cure ailments, and bring wisdom and clarity. These beliefs have contributed to the sapphire's enduring popularity and its presence in various cultural and religious traditions.

Famous Blue Sapphires

The legacy of blue sapphires is enriched by the stories of some of the most famous and historically significant sapphires in existence. These gems not only exemplify the natural beauty of sapphires but also carry with them tales of intrigue, power, and legend.

The Logan Sapphire

One of the most celebrated blue sapphires is the Logan Sapphire, weighing an impressive 422.99 carats. This extraordinary gemstone is named after Mrs. John A. Logan, who donated it to the Smithsonian Institution. Its deep blue color and exceptional clarity make it a masterpiece of nature. The Logan Sapphire is believed to have originated from Sri Lanka, a region renowned for producing some of the finest sapphires in the world.

The Star of India

Another renowned sapphire is the Star of India, a 563.35-carat star sapphire. This gemstone is famous not only for its size but also for the star-like phenomenon known as asterism, which is visible under specific lighting conditions. The Star of India is unique in that it displays stars on both sides of the stone, a rare occurrence. This gemstone, too, has its origins in Sri Lanka and is currently housed in the American Museum of Natural History in New York City.

The Stuart Sapphire

Part of the British Crown Jewels, the Stuart Sapphire is a gemstone steeped in history. This 104-carat sapphire has adorned various royal artifacts, including the Imperial State Crown worn by British monarchs. The Stuart Sapphire has witnessed centuries of British history, from the turmoil of the English Civil War to the splendor of modern coronations. Its journey through time reflects the enduring significance and allure of blue sapphires in royal heritage.

The Blue Belle of Asia

The Blue Belle of Asia, a 392.52-carat sapphire, is one of the most valuable blue sapphires ever discovered. This gem was found in Sri Lanka in 1926 and is renowned for its exceptional size and quality. It was sold at auction in 2014 for over $17 million, setting a record for the highest price ever paid for a sapphire at that time. The Blue Belle of Asia's sale underscores the immense value and desirability of top-quality blue sapphires in the global market.

Conclusion

From their geological formation deep within the earth to their prominent place in human history and culture, blue sapphires are truly remarkable gemstones. Their journey through time, marked by their presence in royal regalia, religious artifacts, and iconic jewelry pieces, showcases their enduring appeal and significance. Understanding the origins and historical context of blue sapphires not only enhances our appreciation of these gems but also connects us to the rich tapestry of human history woven around their captivating beauty.

Chapter 2
The Geological Formation of
Blue Sapphires

Natural Formation

Blue sapphires owe their existence to a combination of rare geological conditions and complex chemical processes that occur deep within the Earth. Understanding these processes not only enhances our appreciation for these precious stones but also underscores their rarity and value.

Formation Process

Blue sapphires are a variety of the mineral corundum, which is composed primarily of aluminum oxide. The blue color in sapphires is primarily due to the presence of trace elements such as titanium and iron. These elements replace some of the aluminum atoms in the crystal lattice, and it is the specific conditions of their incorporation that give rise to the sapphire's blue hue.

The formation of blue sapphires typically occurs in metamorphic and igneous rocks under conditions of high pressure and high temperature. This environment is found deep within the Earth's crust, often in areas with significant tectonic activity.

Metamorphic Rocks

In metamorphic rocks, blue sapphires form through the recrystallization of minerals under intense pressure and heat. These conditions cause the minerals to undergo a physical and chemical transformation, leading to the creation of corundum crystals. The presence of titanium and iron during this process results in the blue coloration of the sapphires.

Igneous Rocks

In igneous rocks, blue sapphires can form in pegmatites and basaltic rocks. Pegmatites are coarse-grained igneous rocks that form during the final stages of magma crystallization. These rocks provide the necessary environment for the slow cooling and crystallization of corundum. Basaltic rocks, on the other hand, are formed from the rapid cooling of lava at the Earth's surface. Sapphires found in basaltic rocks are often brought to the surface by volcanic activity.

Secondary Deposits

In addition to primary deposits found in metamorphic and igneous rocks, blue sapphires can also be found in secondary deposits, such as alluvial deposits. These are formed when the primary rocks weather and erode, releasing the sapphires into riverbeds and other sedimentary environments. The sapphires are then transported and concentrated by the action of water, making them easier to mine.

Major Sapphire Deposits

Blue sapphires are found in various locations around the world, each with unique geological conditions that influence the characteristics of the sapphires produced. The most renowned sapphire deposits are found in Kashmir, Sri Lanka, Myanmar, and Madagascar.

Kashmir

Kashmir sapphires are considered some of the finest in the world, known for their intense, velvety blue color often referred to as "cornflower blue." These sapphires were discovered in the late 19th century in the high-altitude Zanskar range of the Himalayas. The geological conditions in this region, with high-pressure metamorphic rocks, have produced sapphires of exceptional quality and rarity. The limited supply and the remote, challenging location of these mines add to the mystique and value of Kashmir sapphires.

Sri Lanka

Sri Lanka, historically known as Ceylon, has a rich history of sapphire mining that dates back over two millennia. The island's unique geology, characterized by high-grade metamorphic rocks, produces sapphires with a wide range of blue hues, often lighter and brighter than those from other regions. Sri Lankan sapphires are known for their clarity and brilliance, making them highly desirable in the gemstone market.

Myanmar

Myanmar, formerly known as Burma, is renowned for its "Burmese sapphires," which are prized for their rich, deep blue color and excellent transparency. These sapphires are primarily found in the Mogok Stone Tract, a region with a complex geological history involving both igneous and metamorphic processes. The sapphires from Myanmar often exhibit a silky, soft luster that enhances their appeal.

Madagascar

Madagascar is a relatively new player in the sapphire market but has quickly gained a reputation for producing high-quality stones. The island's diverse geology, which includes both igneous and metamorphic rocks, has resulted in a variety of sapphire deposits. Sapphires from Madagascar can range in color from light to deep blue and often rival the quality of those from more established sources. The Ilakaka region in particular has become one of the world's largest sapphire-producing areas.

Mining and Extraction

The process of mining and extracting blue sapphires varies depending on the type of deposit and the geographical location. Both traditional and modern techniques are employed to extract these precious gemstones, with a growing emphasis on sustainable and ethical practices.

Traditional Mining Methods

In regions with alluvial deposits, traditional mining methods are often used. These methods include panning, sluicing, and hand-dug pits. Panning involves washing sediment in a pan to separate sapphires from other materials based on their density. Sluicing uses water to wash sediment through a sluice box, where the heavier sapphires are trapped. Hand-dug pits involve excavating shallow pits in riverbeds or other sedimentary environments to extract sapphires directly from the ground.

Modern Mining Techniques

Modern mining techniques involve more sophisticated machinery and technology to extract sapphires from primary deposits. These methods include open-pit mining and underground mining. Open-pit mining involves removing large quantities of rock and soil to reach sapphire-bearing deposits, while underground mining involves creating tunnels and shafts to access deeper deposits. Both methods require careful planning and management to minimize environmental impact and ensure the safety of workers.

Sustainable and Ethical Practices

The gemstone industry is increasingly recognizing the importance of sustainable and ethical mining practices. This involves minimizing environmental damage, ensuring fair labor practices, and supporting local communities. Efforts to achieve sustainability include restoring mined land, reducing water and energy use, and promoting transparency in the supply chain. Ethical practices also involve certifying gemstones to ensure they are conflict-free and sourced responsibly.

Conclusion

The geological formation of blue sapphires is a fascinating journey that begins deep within the Earth's crust and culminates in the discovery of these precious gems. The unique conditions required for their formation, coupled with the diverse geological settings of major sapphire deposits, contribute to the rarity and value of blue sapphires. Understanding the processes involved in their creation and extraction enhances our appreciation for these extraordinary gemstones and underscores the importance of sustainable and ethical practices in the gemstone industry.

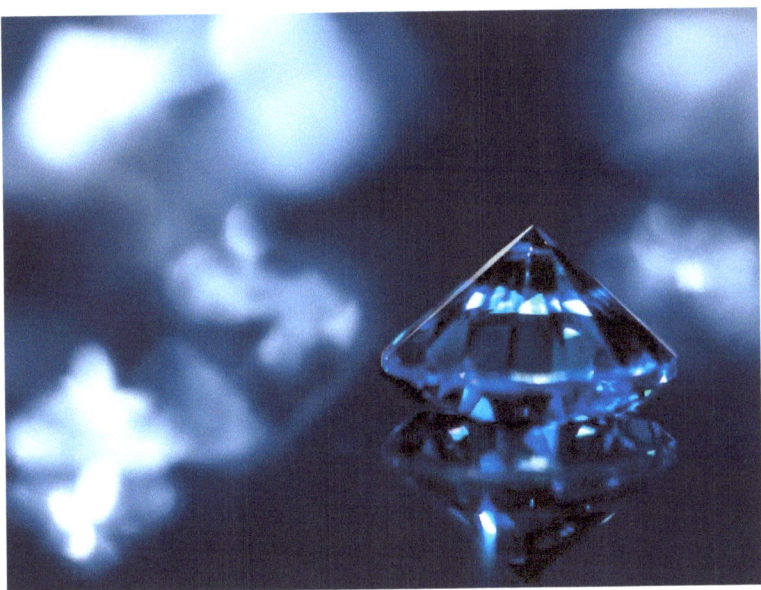

Chapter 3
The Value and Grading of
Blue Sapphires

Factors Affecting Value

The value of blue sapphires is determined by a combination of several key factors, commonly referred to as the four Cs: Color, Clarity, Cut, and Carat weight. Each of these factors plays a crucial role in determining the overall worth of a sapphire, and understanding them is essential for anyone interested in buying, selling, or collecting these precious gemstones.

Color

Color is the most critical factor in valuing blue sapphires. The ideal sapphire color is a vivid, deep blue that is not too dark or too light. The color should be evenly distributed throughout the stone without any zoning (uneven color patches).

Hue:

This refers to the primary color of the sapphire. While the primary hue is blue, the presence of secondary hues such as green or violet can affect the value. Sapphires with a pure blue hue are the most valuable.

Tone:

This refers to the lightness or darkness of the color. Sapphires with a medium to medium-dark tone are generally more desirable.

Saturation:

This refers to the intensity or purity of the color. High saturation, which gives the color a vivid and rich appearance, is preferred.

Clarity

Clarity refers to the presence of internal and external inclusions or blemishes within the sapphire. While inclusions are common in natural sapphires and do not necessarily detract from their value, the type, size, and location of these inclusions can impact the stone's appearance and durability.

Eye-Clean:

Sapphires that appear clear to the naked eye without visible inclusions are more valuable.

Types of Inclusions:

Certain inclusions, like needles of rutile (also known as silk), can enhance the stone's value if they create a desirable effect such as asterism (star effect). However, large or numerous inclusions that reduce transparency or affect the stone's structure can decrease its value.

Cut

The cut of a sapphire influences its brilliance and overall appearance. A well-cut sapphire will reflect light evenly across its surface, enhancing its color and brilliance.

Shape:

Common shapes for blue sapphires include oval, round, cushion, and emerald cuts. The choice of shape often depends on the natural shape of the rough stone and the desired final appearance.

Proportions:

Proper proportions ensure that the sapphire reflects light optimally. Stones that are too deep or too shallow may not exhibit their best color or brilliance.

Symmetry and Polish:

Symmetrical cuts and a high polish enhance the stone's visual appeal and value.

Carat Weight

Carat weight measures the size of the sapphire. Larger sapphires are rarer and therefore more valuable, but size alone does not determine value. A larger sapphire with poor color or clarity will be less valuable than a smaller, high-quality stone.

Price per Carat:

The price per carat increases exponentially with size, especially for high-quality stones.

Size vs. Quality:

While large sapphires are rare, collectors and buyers should consider the balance between size and overall quality.

Grading Systems

Grading systems provide a standardized way to evaluate and compare the quality of sapphires. Various gemological institutes, such as the Gemological Institute of America (GIA), have developed comprehensive grading systems to assess sapphires based on the four Cs.

GIA Grading System

The GIA is one of the most respected gemological laboratories in the world, and its grading reports are widely recognized in the gemstone industry.

Color Grading:

The GIA evaluates sapphires based on hue, tone, and saturation. A detailed description of the color, including any secondary hues, is provided.

Clarity Grading:

The GIA uses a clarity scale that ranges from "internally flawless" to "included," describing the visibility and impact of inclusions under 10x magnification.

Cut Grading:

While the GIA does not provide a specific cut grade for colored stones like it does for diamonds, it assesses the quality of the cut based on the stone's symmetry, proportions, and polish.

Carat Weight:

The exact carat weight is measured and reported.

Other Grading Systems

Other gemological institutes, such as the American Gemological Laboratories (AGL) and the International Gemological Institute (IGI), also provide sapphire grading services. Each institute has its own grading criteria and report formats, but all aim to provide an objective assessment of a sapphire's quality.

Importance of Certification

Certification from a reputable gemological laboratory ensures that a sapphire's quality is accurately represented. It provides buyers with confidence and helps establish the stone's value in the market.

Market Trends

The value of blue sapphires is influenced by various market trends, including economic conditions, fashion trends, and cultural factors. Understanding these trends helps investors and collectors make informed decisions.

Economic Conditions

Economic stability and consumer confidence significantly impact the demand and prices for blue sapphires. During periods of economic growth, luxury goods, including high-quality gemstones, tend to see increased demand. Conversely, economic downturns can lead to decreased demand and lower prices.

Fashion Trends

Fashion trends play a crucial role in shaping consumer preferences for blue sapphires. Celebrity endorsements, red carpet appearances, and designer jewelry collections often drive interest in specific types of sapphires. For example, the popularity of blue sapphire engagement rings surged after Prince William presented Kate Middleton with Princess Diana's sapphire ring.

Cultural and Regional Influences

Cultural significance and regional preferences also affect the market for blue sapphires. In some cultures, sapphires are preferred for their symbolic meanings and are often used in significant life events such as weddings and anniversaries. Regional markets may have distinct preferences for sapphire color and quality, influencing local demand and prices.

Technological Advancements

Advancements in mining, cutting, and treatment technologies have expanded the availability and variety of blue sapphires in the market. These advancements can influence prices by increasing the supply of high-quality stones and introducing new aesthetic enhancements.

Investment Potential

Blue sapphires are increasingly seen as a valuable investment. The limited supply of high-quality natural sapphires, combined with growing global demand, has led to a steady appreciation in their value. Investors consider factors such as provenance, certification, and historical significance when evaluating sapphires as investment assets.

Conclusion

The value and grading of blue sapphires are complex processes influenced by a variety of factors. The four Cs – Color, Clarity, Cut, and Carat weight – provide a framework for evaluating these gemstones, while grading systems from reputable gemological institutes ensure standardization and trust in the market. Understanding market trends and their impact on sapphire values helps buyers, collectors, and investors make informed decisions. As the appreciation for these precious gemstones continues to grow, blue sapphires remain timeless treasures that captivate and inspire.

Chapter 4
Blue Sapphires
In Jewelry and Culture

Symbolism and Meaning

Blue sapphires have been revered throughout history for their striking beauty and profound symbolism. These gemstones are often associated with nobility, truth, sincerity, and faithfulness, making them a popular choice for meaningful jewelry pieces, especially engagement rings.

Nobility and Royalty

Historically, blue sapphires have been linked with nobility and royalty. Their deep blue color was seen as a representation of the heavens, aligning the wearer with divine favor and protection. Kings and queens adorned themselves with sapphires to symbolize their divine right to rule and to invoke wisdom and good judgment.

Truth and Sincerity

In many cultures, blue sapphires are believed to symbolize truth and sincerity. They were thought to protect the wearer from envy and harm, ensuring honesty in relationships. This association makes blue sapphires a popular choice for engagement rings, as they represent the sincerity and faithfulness of the betrothed.

Faithfulness and Loyalty

The enduring color and durability of blue sapphires make them symbols of faithfulness and loyalty. They are often given as gifts to signify long-term commitment and trust, making them ideal for wedding anniversaries and other significant milestones.

Cultural Significance

Blue sapphires hold cultural significance in various traditions around the world. In Hinduism, sapphires are associated with the planet Saturn and are believed to bring good fortune when worn. In ancient Greece and Rome, sapphires were thought to protect against harm and envy. These cultural beliefs continue to influence the way sapphires are valued and worn today.

Iconic Jewelry Pieces

Blue sapphires have been featured in some of the most famous and culturally significant jewelry pieces in history. These iconic pieces have not only showcased the beauty of sapphires but also cemented their place in the cultural lexicon.

Princess Diana's Engagement Ring

One of the most famous blue sapphire jewelry pieces is Princess Diana's engagement ring, which features a stunning 12-carat oval blue sapphire surrounded by 14 diamonds. Chosen by Diana herself from a Garrard jewelry catalog, this ring broke royal tradition by not being a custom-made piece. Following Princess Diana's death, the ring was passed on to her son, Prince William, who gave it to Kate Middleton for their engagement, further solidifying its iconic status.

The Crown Jewels

Blue sapphires have been an integral part of the British Crown Jewels. Notable pieces include the Stuart Sapphire, a 104-carat sapphire set in the Imperial State Crown. The gemstone has a storied history, having been owned by various British monarchs and playing a significant role in royal ceremonies. Its inclusion in the Crown Jewels highlights the long-standing association of sapphires with British royalty.

The Rockefeller Sapphire

The Rockefeller Sapphire is another renowned piece, featuring a 62.02-carat rectangular-cut blue sapphire. Originally from Myanmar, this sapphire was acquired by John D. Rockefeller, Jr. in 1934. It has been set in various jewelry pieces over the years and remains one of the most celebrated sapphires for its size, color, and clarity.

The Star of Bombay

The Star of Bombay is a notable star sapphire, gifted by Douglas Fairbanks to his wife, Mary Pickford. This deep blue, cabochon-cut sapphire weighs 182 carats and displays a striking star pattern due to asterism. The gem, originating from Sri Lanka, is now part of the Smithsonian Institution's collection.

Design and Craftsmanship

The artistry involved in designing and crafting blue sapphire jewelry is a testament to the skill and creativity of jewelers. From selecting the perfect stones to creating intricate settings, the process requires a deep understanding of both the gemstones and the desired aesthetic.

Notable Designers and Jewelers

Several renowned designers and jewelers have mastered the art of incorporating blue sapphires into their creations, elevating these gemstones to new heights of beauty and elegance.

Cartier:

Known for its luxurious and innovative designs, Cartier has created numerous iconic sapphire jewelry pieces. The Cartier Halo Tiara, which features a stunning array of sapphires, is one such masterpiece.

Harry Winston:

Often referred to as the "King of Diamonds," Harry Winston also has a legacy of working with blue sapphires. His designs are characterized by their sophistication and impeccable craftsmanship.

Bvlgari:

This Italian luxury brand is famous for its bold and colorful designs. Bvlgari's use of blue sapphires often incorporates vibrant contrasts with other gemstones, creating striking and contemporary pieces.

Design Techniques

Creating blue sapphire jewelry involves several design techniques that enhance the gemstone's natural beauty.

Halo Settings:

Halo settings surround the central sapphire with a ring of smaller diamonds, amplifying the sapphire's brilliance and making it appear larger.

Pavé Settings:

Pavé settings involve setting small diamonds closely together around the sapphire, adding extra sparkle and highlighting the main gemstone.

Bezel Settings:

Bezel settings encase the sapphire in a metal rim, offering a modern and secure setting that protects the gemstone while showcasing its color and clarity.

Crafting Process

The crafting process of blue sapphire jewelry involves multiple stages, from initial design to final polishing.

Stone Selection:

Selecting the right sapphires is crucial. Jewelers look for stones with optimal color, clarity, and cut.

Design Concept:

Designers create sketches and models to visualize the final piece. This stage involves deciding on the overall aesthetic, including the choice of metal and additional gemstones.

Setting and Mounting:

The selected sapphires are carefully set into their mounts. This stage requires precision and expertise to ensure the stones are secure and displayed to their best advantage.

Polishing and Finishing:

The final piece is polished to enhance its shine and finish. This step brings out the full beauty of the gemstones and the metalwork.

Conclusion

Blue sapphires have a rich cultural heritage and are deeply embedded in the traditions and symbolism of many societies. From their significance in royal regalia to their status as symbols of truth and loyalty, these gemstones have captivated the imagination of people worldwide. Iconic jewelry pieces featuring blue sapphires continue to inspire awe and admiration, while the artistry and craftsmanship involved in creating sapphire jewelry highlight the enduring appeal of these precious stones. Understanding the cultural and artistic context of blue sapphires enhances our appreciation for these remarkable gems and their place in the world of fine jewelry.

Chapter 5

The Future of Blue Sapphires

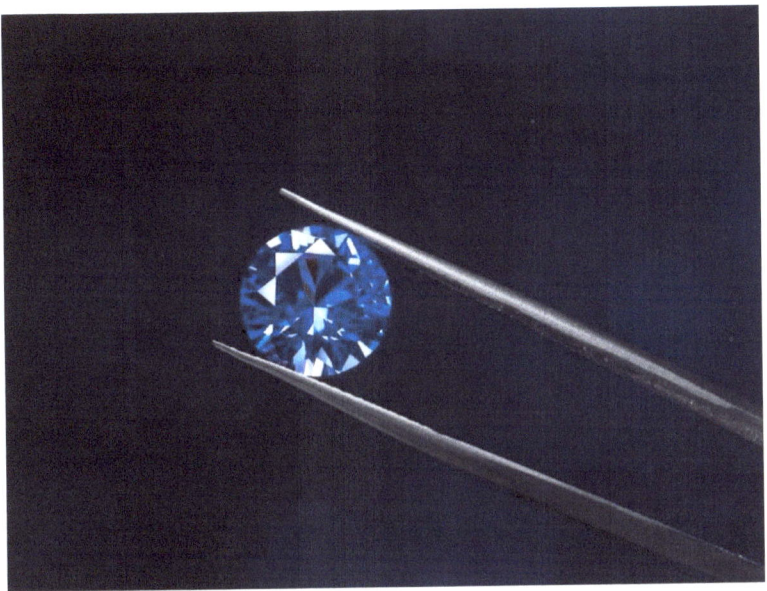

Sustainability and Ethical Practices

As awareness of environmental and ethical issues grows, the gemstone industry is increasingly focused on sustainable and ethical practices. This shift is critical for the future of blue sapphires, ensuring that their beauty is matched by responsible sourcing and production.

Sustainable Mining Practices

Traditional mining practices have often had detrimental impacts on the environment, leading to habitat destruction, water pollution, and soil erosion. In response, the gemstone industry is adopting more sustainable mining practices to mitigate these effects.

Eco-Friendly Techniques:

Modern mining operations are implementing eco-friendly techniques such as low-impact mining and rehabilitation of mined land. These practices help reduce the environmental footprint of sapphire extraction.

Waste Management:

Proper waste management systems are being established to handle mining byproducts responsibly, preventing contamination of local water sources and soil.

Energy Efficiency:

The use of renewable energy sources and energy-efficient machinery in mining operations reduces carbon emissions and promotes sustainability.

Ethical Sourcing and Fair Trade

The ethical sourcing of sapphires is paramount to ensure that the gemstones are not linked to human rights abuses or conflict financing. Fair trade practices in the gemstone industry aim to improve the livelihoods of miners and their communities.

Fair Wages and Working Conditions:

Ensuring that miners receive fair wages and work in safe conditions is a key aspect of ethical sourcing. This involves adhering to international labor standards and providing necessary safety equipment and training.

Community Development:

Mining companies are increasingly investing in local communities, funding education, healthcare, and infrastructure projects to promote long-term development.

Transparency and Traceability:

Establishing transparent supply chains and traceability systems helps verify the ethical origins of sapphires. Consumers are increasingly demanding information about the provenance of their gemstones, and technology like blockchain is being used to track sapphires from mine to market.

Technological Advancements

Advancements in technology are transforming the gemstone industry, from mining and cutting to treatment processes. These innovations enhance the quality and availability of blue sapphires while also addressing some of the challenges faced by the industry.

Advanced Mining Techniques

Innovative mining techniques are improving the efficiency and sustainability of sapphire extraction.

Remote Sensing and Exploration:

Technologies like remote sensing and satellite imaging help identify sapphire deposits with minimal environmental disruption. These methods allow for more precise targeting of mining areas, reducing the need for extensive and invasive exploration.

Automation and Robotics:

The use of automation and robotics in mining operations increases efficiency and safety. Automated machinery can perform tasks that are hazardous to human workers, reducing the risk of accidents.

Cutting and Treatment Innovations

Technological advancements in cutting and treatment processes enhance the beauty and durability of blue sapphires.

Precision Cutting:

Computer-aided design (CAD) and precision cutting technologies allow for more accurate and consistent cuts, maximizing the brilliance and value of sapphires.

Heat Treatment:

Heat treatment is a common practice used to enhance the color and clarity of sapphires. Advances in treatment techniques ensure more uniform and stable results, improving the overall quality of treated stones.

Synthetic Sapphires

Synthetic sapphires, created in laboratories, offer an alternative to natural sapphires. These stones are chemically identical to their natural counterparts and can be produced with fewer environmental impacts.

Production Methods:

Methods such as the Verneuil process, Czochralski process, and hydrothermal synthesis are used to create synthetic sapphires. These techniques allow for the production of high-quality sapphires with controlled color and clarity.

Market Position:

While synthetic sapphires are often less expensive than natural stones, they are gaining acceptance in the market for their ethical and environmental benefits. They provide an affordable option for consumers who seek the beauty of sapphires without the ethical concerns associated with natural mining.

Investment and Collecting

Blue sapphires are increasingly seen as valuable investments, and building a collection requires careful consideration of several factors to ensure that these precious gemstones retain their value over time.

Factors to Consider

When investing in blue sapphires, several key factors must be considered to make informed decisions and maximize returns.

Quality:

As discussed in previous chapters, the four Cs (Color, Clarity, Cut, and Carat weight) are critical in determining the value of a sapphire. High-quality stones with exceptional color and clarity are more likely to appreciate in value.

Certification:

Obtaining certification from reputable gemological laboratories, such as the GIA or AGL, provides assurance of a sapphire's quality and authenticity. Certified stones are more attractive to buyers and collectors.

Provenance:

The history and origin of a sapphire can significantly influence its value. Sapphires with notable provenance, such as those from famous mines or with historical significance, are highly sought after.

Building a Collection

Building a sapphire collection involves strategic planning and knowledge of the market.

Diversification:

Diversifying a collection by acquiring sapphires from different regions and with varying characteristics can mitigate risk and increase the potential for appreciation.

Market Trends:

Keeping abreast of market trends and understanding the factors driving demand can help in making timely and profitable investment decisions.

Professional Guidance:

Consulting with gemologists, appraisers, and dealers can provide valuable insights and help navigate the complexities of the sapphire market.

Care and Maintenance

Proper care and maintenance are essential to preserve the value and beauty of blue sapphires.

Cleaning:

Sapphires should be cleaned regularly with mild soap and water. Ultrasonic and steam cleaners can also be used for thorough cleaning, but care should be taken with treated stones.

Storage:

Sapphires should be stored separately from other gemstones to avoid scratches. Soft pouches or lined jewelry boxes are ideal for storage.

Regular Inspections:

Regular inspections by a professional jeweler can help detect any issues, such as loose settings or damage, and ensure that sapphires remain in optimal condition.

Conclusion

The future of blue sapphires is shaped by sustainability and ethical practices, technological advancements, and their growing appeal as investment assets. The gemstone industry is evolving to meet the demands of a conscientious market, ensuring that the beauty of blue sapphires is matched by responsible sourcing and production. As innovations continue to enhance the quality and availability of sapphires, and as their status as valuable investments solidifies, blue sapphires will continue to captivate and inspire future generations. By understanding the complexities of the market and the importance of ethical practices, collectors and investors can appreciate the true value of these precious gemstones and contribute to a more sustainable and equitable industry.

The End!

www.ingramcontent.com/pod-product-compliance
Lightning Source LLC
Chambersburg PA
CBHW040758240526
45474CB00008B/104